The Midland & South Western Junction Railway
Images from the Transport Treasury collection

Compiled by Jeffery Grayer

© Images and design Transport Treasury 2022 Text Jeffery Grayer

ISBN 978-1-913893-27-9

First published in 2022 by Transport Treasury Publishing Limited. 16 Highworth Close, High Wycombe, HP13 7PJ

Totem Publishing an imprint of Transport Treasury Publishing.

**The copyright holders hereby give notice that all rights to this work are reserved.
Aside from brief passages for the purpose of review, no part of this work may be reproduced, copied by electronic or other means, or otherwise stored in any information storage and retrieval system without written permission from the Publisher. This includes the illustrations herein which shall remain the copyright of the copyright holder.**

www.ttpublishing.co.uk

'*The Midland & South Western Junction Railway*' is one of a series of books on specialist transport subjects published in strictly limited numbers and produced under the Totem Publishing imprint using material only available at The Transport Treasury.

Front Cover: The principal train of the day, the Cheltenham - Southampton through service, arrives at Swindon Town apparently with steam to spare as the signalman readies himself to receive the single line token. Maunsell moguls were the usual motive power for this train and U Class No. 31613 based at Eastleigh shed was a regular performer on this duty. *H1805*

Back cover: Informative platform sign mounted beneath the footbridge at Swindon Town station photographed on 23rd October 1959 advising passengers of the plethora of destinations available at the "other" railway in the town should they wish to change on to the "Bunk" or the "Dodger" as the shuttle service between the former GW and M&SWJR stations was known locally. *H1138*

Frontispiece: Following diversion of services from the High Level to the Low Level station at Savernake effective from 15th September 1958 tracks through the former were utilised for wagon storage as this view taken on 19th September 1959 reveals. Although this closure was billed as temporary, citing weak underbridges crossing the Kennet & Avon canal and the Western mainline near Wolfhall Junction, this like so many other "temporary" closures turned out to be permanent and by the spring of 1960 tracks had been lifted through the High Level station. *LRF4403*

Introduction

Closing over sixty years ago, in September 1961, there is little left today of the infrastructure of the M&SWJR with just the odd bridge or embankment surviving to show the former line of this 61 mile cross-country route from Andover to Andoversford. There are however a couple of notable exceptions, as the Swindon & Cricklade Railway Society is seeking to recreate a flavour of the old route on their 2½ mile preserved section, and a single track from Andover as far as Ludgershall remains in situ. This used to carry MOD military traffic and although regular services ceased in 2015 there is the occasional movement with one of the most recent being in March 2021 when a Class 66 brought in just one wagon from MOD Kineton. The track seems to be maintained in immaculate condition and there has even been talk in recent years of reinstating a passenger service between Ludgershall and Andover.

Following its undoubted usefulness during the two world wars the line's fortunes experienced a rapid decline reflected in the curtailment of services during the 1950s. In 1955 the branch passenger service between Ludgershall and Tidworth was withdrawn followed in 1958 by savage reductions in train services which left just one train a day in each direction running the whole length of the line. This was a far cry from the line's busiest days during WW1 when in 1915 for example there were seven southbound services traversing the whole route including a Boat Express serving the American and Cape lines and through trains from both Nottingham and Manchester. This level of service was subsequently reduced to four through trains and by 1953 the figure was down to three. Following the 1958 pruning there remained just a single daily return service over the whole line, with a further two services between Swindon and Andover together with two short workings (SX) from Cirencester to Swindon, a short working from Swindon Town to Marlborough (SO) and two workings (one SX) from Swindon Town to the next station south at Chiseldon. In a northerly direction, in addition to the solitary Southampton – Cheltenham train, there was one service from Andover to Savernake, one from Andover to Swindon plus two workings from Swindon to Cirencester, three short workings from Chiseldon to Swindon (two SX) and two from Savernake to Swindon (one SO). Surprisingly there were two services which continued to run on Sundays in each direction between Andover and Swindon. This meant that there was just one train a day in each direction on the section between Andoversford and Cirencester.

Additionally the Savernake to Marlborough shuttle provided some half dozen workings, one being Saturday only, daily each way connecting the Wiltshire market town with the WR mainline from Reading to Westbury. At the Cheltenham end the traditional link with the Midland at Lansdown Junction was severed and services diverted to the ex GWR terminus at St. James thereby making connection with Midland route from Bristol to Birmingham more difficult for passengers. In the same year the High Level station at Savernake was closed, ostensibly as a temporary measure which turned out to be permanent, and all services diverted to run via the Low Level station. Following closure in September 1961 a remnant of the passenger service continued for a while with a morning and evening workmen's service from Cirencester to Swindon honouring a long established agreement to cater for workers redeployed to Swindon Works following closure of the M&SWJR works in Cirencester soon after the grouping. Elsewhere on the line goods traffic continued on isolated sections of track, from Swindon Town to Cirencester, from Andover to Ludgershall, and from Swindon (WR) to Swindon Town and to Moredon power station. The Marlborough to Savernake link continued to be served by goods trains and the occasional school special until September 1964. The final goods closure, that from Swindon (WR) to Swindon Town, occurred as late as 1972.

Looking at the results of a passenger survey undertaken by BR during week ending 19th November 1960, it was hardly surprising that the line did not pay with the majority of stations recording average daily passenger numbers in single figures. The only locations recording numbers in excess of ten were Swindon Town, Chiseldon, Marlborough, Savernake, Collingbourne, Ludgershall and Andover and even then the highest daily average was only 89 joining and a similar number alighting at Marlborough the majority of whom were no doubt only making the short trip to and from Savernake to connect with the mainline. In March 1961 BR wrote to local authorities advising them of the plan to close the line and requesting that they lodge any objections with the local TUCC (Transport Users Consultative Committee). The outcome, given the parlous state of the line's finances and general lack of patronage reflecting the inadequate train service offered, was a foregone conclusion and the line duly closed on and from 11th September 1961.

With its infrequent service of trains, some stations only having the luxury of one train a day in each direction, coverage of the M&SWJR by contemporary photographers, who often relied on public transport, was somewhat limited but sufficient images have been located within the vast Transport Treasury archive, sourced from several different collections, to enable this publication hopefully to capture something of the essence of the line's operations during its last years of operation. Beginning with a look at the route we move on to focus on perhaps its most interesting working the through Southampton to Cheltenham service. The former headquarters of the line in Swindon and the shuttle operation running between the Berks & Hants route at Savernake and Marlborough are also covered and we conclude with a look at the various types of motive power to be seen on the line, including a couple of views from 90 years ago, leading to that final weekend of services and specials.

Jeffery Grayer 2022

The Route

The moment of token exchange at Red Post Junction to the west of Andover is captured by the photographer on board the afternoon Swindon Town to Andover service hauled by U Class No. 31795. This wartime "utility" signalbox was opened in 1943 when a junction was made with the LSWR double track mainline although the original MSWJR line continued to parallel the main line right into the Hampshire town. The signalbox closed in September 1963.
H1827

This June 1966 view of a rather dilapidated Weyhill station reveals that the former double track on the section from Andover, which was put in as a wartime measure during September 1943, has been singled. This was carried out in August 1961 but notice that a couple of platform barrows remain on the platform in the expectation of any goods traffic offering, a facility which was not withdrawn here until December 1969. This was a far cry from the days of the busy Weyhill Fair which originally took place three times in the year, in April for trading cattle, July for selling lambs and in October for hops. The goods yard would be full to capacity at fair time with nearly 200 wagons necessitating the employment of no less than three shunting horses. The fair was immortalised by Thomas Hardy as Weydon Priors in his novel "The Mayor of Casterbridge" where the protagonist Michael Henchard sells his wife. *RCR17966*

In spite of appearances Weyhill signalbox was still operational at the time of this view taken on 9th July 1956. This 20 lever box to a standard Gloucester Wagon Co. design was to close consequent upon the singling of the line to Andover. Andover Museum now houses the signalbox sign together with one of the Weyhill station signs. Double track between here and Ludgershall had been instituted in August 1900. *R507*

The impressive station of Ludgershall, one of only three on the route to boast a footbridge which in this instance dated from 1902, is seen in this view looking towards Andover on 8th July 1956. The size of the station and the length of the platforms was a reflection of the importance of military traffic both here and at nearby Tidworth enabling the speedy handling of both men and horses. This led to a rather bleak and windswept station not particularly welcoming to the ordinary traveller. *R7495*

This close up of the small shelter on the northbound platform providing some small comfort for waiting passengers and still advertising the long closed connection for Tidworth was photographed on 2nd July 1960 almost five years after withdrawal of the branch service. A military Land Rover is parked on the platform whilst one of the posters inside the wooden shelter advertises the delights of Burnham-on-Sea. *NF044-3*

This view dating from 9th July 1956 was taken from the signalbox at Ludgershall which was erected in 1901 to an LSWR design and closed in 1963. It contained 40 levers and afforded a good view of the former branch platform used by the Tidworth service in the foreground. The track leading north under the bridge which carried the A3026 Tidworth road had been doubled as far as Collingbourne in September 1901 with the section onwards to Grafton being so treated the following year. The overbridge still crosses the remaining tracks today but the main road now follows a revised course a little to the north crossing the line by a newer structure. *R7512*

Taken from a southbound service in July 1960 this view illustrates the junction of the Tidworth branch curving off to the left. Although the majority of the branch was single track there was originally a double track section extending from Ludgershall station as far as Perham signal cabin but this was singled in 1955 when the cabin closed although some of its levers were retained for use as a ground frame. The Goods shed, outside of which stands a solitary wagon, is prominent in the area between the Tidworth line and the mainline. *NF044-26*

Collingbourne station presents a neat appearance in this image taken on 8th July 1956. The proximity of the village of Collingbourne Ducis is apparent and consequently the station was quite well used and even as late as November 1960, when a survey of passenger usage was undertaken by BR, some 17 passengers joined and 11 alighted on average per day which may not seem much but one has to remember that by this time Collingbourne only saw 3 services daily each way. The difference in platform levels apparent in this view indicates some extension to platform length had taken place, the platforms being lit by paraffin vapour lamps suspended from tall concrete posts. A light railway was at one time proposed from Collingbourne to Netheravon where one of the promoters of the M&SWJR owned property but the scheme failed. The Ducis part of the village name originates from the fact that John of Gaunt inherited the manor of Collingbourne and he later became Duke of Lancaster hence the village was known as Collingbourne Dukes or Ducis. *R7481*

The pastoral countryside south of Collingbourne Kingston Halt sees the passage of Churchward 4300 Class 2-6-0 No. 6320 on the 8th July 1956 with an Andover service consisting of two coaches and a van. The small halt which opened in 1932 can just be glimpsed in the background before the overbridge carrying a minor road to the nearby hamlet of Brunton. *R7480*

London and South Western Ry.
787
From WATERLOO
TO
COLLINGBOURNE
Via Andover Junction.

The halt at Collingbourne Kingston is seen in detail on 23rd October 1959. It consisted of two short wooden platforms lit by electric lighting with corrugated iron huts on each. No board crossing was provided as access to both platforms was possible from the adjacent road bridge. Tickets were available from a house on the main A338 road which ran through the nearby village and although by the time of the BR survey in 1960 only some 7 passengers were joining and 6 alighting daily on average from trains here back in the 1930s soon after opening some 1500 tickets had been sold annually. No trace of the halt now remains although the overbridge still spans the grassy trackbed. *H1135*

Quite a reasonable crowd of waiting passengers, at least by M&SWJR standards, greets the arrival of U Class mogul No. 31639 as it drifts into Grafton & Burbage station on 15th March 1958. The milepost on the slightly shorter up or northbound platform indicates 12½ miles from Red Post Junction and SR upper quadrant signals are in use following the transfer of the line southwards to the SR on 2nd April 1950 although this section reverted back to WR control on 1st. February 1958 until closure. *RCR11526*

By July 1960, the date of this view, Grafton South Junction which had originally been named Wolfhall Junction in M&SWJR days but was renamed in 1933 to avoid confusion with the similarly named GWR box, had lost its connecting 1905 south to east chord which had only been used intermittently by mainly military and excursion traffic and which had been taken out of use on 5th May 1957. This chord gave access to the Berks & Hants route to Newbury and Reading. The box also controlled the junction of the lines to both High and Low level stations at Savernake and it remained open until closure of the line. The name of the signalbox recalls its proximity to the site of Wolf Hall the home of Jane Seymour and which has come to prominence again recently in Hilary Mantel's Booker prize winning series of novels about Thomas Cromwell. *NF044-24*

And so we come to one of the most intriguing locations on the line – Savernake. The history of the two stations is too complex to go into detail here but suffice it to say that this is the view from a train that has called at the M&SWJR station known as High Level, on 21st August 1957. By this date the down platform seen here and passing loop, which had been taken out of service for passenger traffic back in 1933, was no longer used by the remaining services and it would only be another 13 months before all trains were diverted to the WR Low Level station 200 yards to the south. This of course allowed far easier interchange with services on the Berks & Hants line where one could change for destinations as diverse as Devizes, Trowbridge, Bath, Bristol, Yeovil, Dorchester, Weymouth, Newbury, Reading and London Paddington.
LRF2936

This view of the exterior of the High Level station at Savernake dating from 19th September 1959 reveals not only the main station building but also the water tower dating from 1898, signalbox, which was another product of the Gloucester Wagon Co., and, on the opposite platform, the private waiting room provided for the Marquess of Ailesbury who had leased land in Savernake Forest to the railway for a rental of £450 p.a. plus an agreement to provide a station at Savernake "adjacent to, and at least equal to, the existing station of the GWR". The motto of the Marquess was rather appropriately "Think and Thank". The signalbox formerly contained an 18 lever frame but after ceasing to function as a block post in 1933 it was retained merely as a ground frame known as "High Level Middle Ground Frame". *LRF4404*

This seldom seen viewpoint looking towards Marlborough from the road bridge by the High Level station was also taken on 19th September 1959. It gives the lie to previously published details of track removal between here and Marlborough in that it was in fact the up line rather than the down line from Marlborough that was retained. This view clearly shows that the former down line has been slewed into the up line with track having been recovered leaving a remaining single line which continued all the way to Marlborough. This then became double using the former down line after the junction with the new connection from the original line from Savernake Low Level most of which had been taken out of use following the track alterations of 1933. Part of the original GWR line from Savernake Low Level to Marlborough was retained at the Marlborough end to act as a headshunt to permit access to the former High Level station and yard at Marlborough.
LRF4406

By July 1960 the scene at the High Level station was much transformed by the removal of all tracks although this does give us a closer view of the signalbox and associated water tower. Today the main station building has been converted into an attractive dwelling. *NF044-21*

Quite a contrast at the other Savernake station where Modified Hall 4-6-0 No. 7922 "Salford Hall" tears through the Low Level platforms on an unrecorded date carrying reporting number A92 which, in 1960 for example, referred to the 13:45 SO Paignton to Paddington service due into London at 18:15. As the large running In board testifies this is the junction for Marlborough and a two coach set waits in the bay platform to provide the next service. At this time there were five services each way between the junction and Marlborough on Mondays-Fridays, six on Saturdays and two on Sundays with a journey time of just ten minutes. *H1824*

At a rather more sedate pace 5700 Class pannier tank No. 4697 comes round the curve from Marlborough in July 1960 with an Andover service. Notice the 5 mph speed restriction applying to services leaving the bay platform whilst in the distance can be seen Savernake West signalbox which was not finally taken out of service until November 1978. *NF044-23*

Savernake Low Level witnesses the arrival of U Class No. 31618 with the 4:52pm Swindon to Andover service on 19th. September 1959. The attractive patterned brickwork of the station building is apparent and, following closure of the M&SWJR route in September 1961, the station continued to be served by freight trains until May 1964 and by a sparse stopping passenger service on the Berks & Hants route until closure came in April 1966.
LRF4409

A pannier tank hauled train approaches the northern portal of the 648 yard tunnel the construction of which was necessitated by the desire of the M&SWJR to take a more direct route from Savernake to Marlborough than that pursued by the GWR. This decision came at a cost with the tunnel requiring major repairs on two occasions, in 1925 when relining took place and in 1944 when further partial relining was necessary, and with falling chalk, evidence of which is seen on the left, being a constant danger on the steep sided 70 foot deep cuttings. The track on the left led to Savernake High Level station after passing over a summit at nearly 600 feet above sea level on the high plateau to the south. The tunnel is today in a poor state and currently houses a bat sanctuary. H1819

Marlborough, seen here also on 19th September 1959, was one of the more important stations on the line and following the closure of the separate GWR station in March 1933 played host not only to M&SWJR trains but also to the shuttle service to Savernake. The substantial girders of bridge No. 50 in the foreground crossed the main A346 road from Salisbury into the town. The Refreshment Room signposted on the up platform was opened in 1884 and continued to serve as a local hostelry complete with dart board for some time after the closure of the station to passengers in 1961. The licensee at the time of closure, one Harold Trotman, told a local reporter that he was so busy on the final day that "he wished they would close the railway every day". Indeed in later years the exterior was painted white with the words "Teas & Snacks, Licensed Bar" emblazoned on the roof in large letters looking rather incongruous next to the boarded up trackless station. *LRF4397*

The rarely photographed exterior of Marlborough station reveals that the refreshment room-cum-pub appears to be doing good business judging by the number of beer crates awaiting collection. A BR parcels lorry seen in front of the substantial goods shed also adds interest to this 1959 scene. The chimneys of the main station building, which was larger than many such buildings on the line reflecting its relative importance, had when first constructed been very tall but following concerns raised in 1910 that they had become somewhat "shaky" these were subsequently reduced in height to their rather truncated state seen here. *LRF4396*

The former GWR terminus is seen in this 10th September 1961 view during the final weekend of services on the adjacent M&SWJR whose own Marlborough station can be seen at a lower level on the far right with the carriages of a train standing at the up platform. Freight and school traffic continued to be handled at Marlborough until May 1964 after which it only handled coal until final closure to all traffic came in September 1964. The signalbox seen on the right was resited here from its original position at the other end of the down platform in 1933 when the rationalisation of lines between Marlborough and Wolfhall Junction east of Savernake took place.

A remarkable survivor at the GWR station was the former single road engine shed, closed 26 years before, with its associated water tower seen here in September 1959. A couple of enthusiasts wander about the site and peer through cracks in the shed doors whilst the laundry on the adjacent house's washing line would in all probability remain unsullied by any locomotive smuts. *LRF6579/LRF6577*

Ogbourne station is bathed in summer sunshine in this July 1960 shot taken from a northbound service. After 1952 the passing loop here was signalled for two way working although trains seldom crossed here the one exception being a Saturdays only service from Swindon (WR) to Marlborough which crossed with a northbound local from Marlborough to Swindon. The raised height of part of the platform evident in this view was done to facilitate the loading of milk churns and cattle, the latter housed in the rather dilapidated pens seen on the far left. The village was situated on the main road between Marlborough and Swindon and thus benefitted from a much more frequent bus service meaning that the station was little used in later years. Following closure of the line in 1961 the final ignominy came when a by-pass for the village was constructed on the former track bed. NF 045-1

Chiseldon Camp halt lay just 1¼ miles to the south of Chiseldon and consisted of a single sleeper built platform that was provided with the luxury of electric lighting. As a wartime measure a loop and siding were provided from 1943 until 1950 to the north of the halt. The final army unit to be stationed at the camp departed in 1962 and the site was left virtually empty until 1974 at which time demolition began. One of the few traces of the camp now remaining is the old railway halt platform whilst the Department of Transport's Heavy Goods Vehicle Driver Test Centre is located on the site of the former camp hospital. *H1808*

Chiseldon station seen here looking south on 23rd October 1959 was quite well used especially by commuters and shoppers going into Swindon for the day and by schoolchildren as train timings were well suited to office and school hours and indeed two services daily from Swindon terminated here. The separate Goods Office which had been added during the GWR era can be seen on the up platform and bi-directional running was in operation from 1952 when the signalbox here could be switched out. At one time the area's famous racing stables had provided much horsebox traffic for the railway. *H1137*

Evidence of the popularity of the train service at Chiseldon is provided by this 3rd September 1960 view of passengers boarding a Swindon bound service hauled by 5700 Class pannier tank No. 9721 sporting an 82C (Swindon) shedplate. The adjacent Elm Tree pub with its porch seen on the left still stands guard today although it now looks out on a grassy area where the old station could once be found. *H1796*

The third station on the line to boast a footbridge, the others being Ludgershall as we have seen and a long demolished one at Savernake High Level, is prominent in this view of a virtually deserted Swindon Town looking towards Cheltenham on a rather damp 23rd October 1959. The signalbox designated "A" box was to an LSWR design having replaced the 1881 original in 1905 and contained 17 levers. The station had been enlarged, also in 1904-5, with the creation of an island platform on the up side and the provision of an adjacent road that led to the turntable just visible on the far left of this view. *H1140*

A busier scene at Swindon Town reveals pannier tank No. 9740 waiting to attach a van on to an incoming train hauled by U Class No. 31793 on 15th August 1959. The station platforms were over 500 feet long, an overly generous provision in the latter days of two or three coach trains. Water cranes were provided at both ends of the station and in the far distance can be seen the 150 foot long Goods Shed which contained an internal loading platform and 5 ton crane. General goods traffic ceased from May 1964 but coal continued to be dealt with until November 1966 and oil trains to the nearby Esso depot ran until 1968.
H964

This view looking north west shows an unidentified pannier tank at the head of a local to Swindon WR in the bay platform on 20th April 1960. The substantial station buildings were not demolished until January 1968 following damage by fire and during 1970/1 the site was razed to the ground but stone traffic, from Mendip quarries in connection with construction of the nearby M4, saw a return of freight trains briefly during this period. Track was not finally taken up between Swindon Town and the former Moredon power station until 1978. The site has now been redeveloped for housing and light industry. *NF030*

This eve of closure view of part of the layout at Rushey Platt junction dates from 8th September 1961. The original station building seen on the left was situated on the spur that connected the M&SWJR with the WR mainline, the main M&SWJR line running behind the signalbox seen on the right. The signalbox was to an LSWR design opened in 1917 and did not close until June 1965 having a 30 lever frame. The low level platforms had closed back in 1905 whilst the high level platforms served in later years merely as staff platforms although goods, mainly milk, continued to be handled here until freight services were withdrawn between Moredon and Cirencester in 1964. *H3195*

A closer view of the signalbox at Rushey Platt taken in July 1960 also reveals the 7:50 pm from Savernake descending behind the box taking the spur line that led to Swindon WR station. This train was only extended from Swindon Town to Swindon WR on Saturdays. The tablet exchanging apparatus with their adjacent illumination in the form of oil lamps on poles, are well shown and attached to the signalbox window is a rather insignificant notice advising that speed should not exceed 15 mph. A token was required for the single line onwards to Swindon Town, the double track apparent here was merely a passing loop. *NF045-7*

A lone passenger waits at Cricklade station at some time during April 1960. The BR survey of 1960 reveals the average number of passengers joining trains here amounted to the grand total of 1 with none alighting – this Thameside town had long ago deserted its railway for the more convenient buses. Like many stations on the line by this time Cricklade was beginning to look rather unloved – a symptom of the lack of interest shown in the route by the Western Region – and even the canopy columns retain some of their wartime white paint indicating that repainting has not been a priority. The end of the separate loading bay can just be glimpsed on the far right. Milk was at one time a major traffic here and a clutch of four empty milk bottles near the dustbin are a sad echo of this traffic lost to the roads. *NF030-24*

A barely legible running in board proclaims this to be Cricklade, this view having been taken on 26th July 1961. The signalbox here which contained 14 levers remained in service until the end of passenger services but was subsequently downgraded to a ground frame until 1st July 1963 when freight services were withdrawn. The crossing loop here had been extended northwards as part of wartime improvements in 1942. Today the station site and much of the route west of the town lies buried under a realigned B4040 so any plan of the preservationists to reach the town will have to include a new site for a station here. *H2821*

Taken from the first carriage behind the distinctive outline of a Maunsell mogul tender on 19th May 1959 this is a view of South Cerney down platform looking south. The nameboard seems to be in better condition than that at Cricklade and the signalman's motor cycle is propped up against the box in time honoured fashion whilst the man himself can be seen at the end of the platform ready to exchange tokens with the driver of the through train from Cheltenham to Southampton. The signalbox with a 28 lever frame dates from 1942 when the loop here was lengthened as a wartime measure and like Cricklade it became a ground frame after closure to passengers until May 1964 although freight traffic, much of it being gravel taken from the surrounding pits, had ceased here in July 1963.
LRF4390

The up platform of South Cerney is glimpsed through the arches of the adjacent road bridge as the Southampton train gets away at just before 3pm on the last Friday of operations, the 8th September 1961. The route between Marlborough and Cirencester was never doubled but a passing loop had been installed here in 1900. With only three services daily each way, two of these going to or starting from Cirencester, average passenger numbers were as to be expected virtually nil by this time although undoubtedly some of the locals turned out to take a last ride over this final weekend. *H3194*

Cirencester was certainly one of the larger settlements on the route and in times past had been served by several services daily. However, by the date of this view, 19th September 1959, the town's Watermoor station had seen its train service shrink to just one northbound and three southbound services daily. The truth was that most locals preferred to catch the shuttle service from the Town station to Kemble where they could change for Swindon or Gloucester and Cheltenham. The signalbox was somewhat unusually labelled "Cirencester Station Cabin" and contained 19 levers although its days were numbered being closed in August the following year. The red flag positioned in the middle of the up track may indicate some trackwork going on but following damage to an underbridge to the north of the station the up platform was no longer used and by early 1961 the track had been taken up as shown in a subsequent image. *LRF4389*

The attractive Cotswold stone construction of the main building is shown to advantage in this exterior view of Cirencester Watermoor dating from July 1960. A full length canopy was provided and Cirencester became an important administrative centre for the M&SWJR with Sam Fay, the driving force behind the ailing company's reversal of fortune, basing himself here being in part a reflection of the fact that the company's locomotive works was also located here. Sadly a roundabout and new roads occupy the site today there being no trace of the station remaining. *NF047-15*

A view of Cirencester station looking south reveals that the up line is no more with just sleepers marking its former position. It was unfortunate that the sole remaining down line had rather inferior waiting facilities on its platform but with only one northbound train and three southbound services perhaps this hardly mattered by this stage. The stone water tower which supported an overly large replacement tank was situated next to the small building clad in corrugated iron which had previously been used by various staff including the stationmaster, locomotive superintendent and the civil engineer.

Beyond the grass grown tracks of the former goods yard at Cirencester lay the former Works complex seen here on 26th July 1961. This had been established here by the M&SWJR in 1895. Employing some 80 men in the early 1920s the Works closed in 1925 with many staff subsequently transferring to Swindon Works. The main locomotive shop with the broad arched entrance and circular high level window contained an overhead travelling crane. By the time of this view much of the area had become a scrapyard as evidenced by the piles of metal, old vehicles and tyres littering the site. The spire of Holy Trinity church Watermoor, designed by the noted architect Sir Giles Gilbert Scott, is prominent in the background. *H2819*

Considering the isolated position of lonely Foss Cross station on windswept uplands without much in the way of local habitation, the provision of facilities here seemed overly generous but can be largely explained by the fact that this was a useful location for a crossing place between Withington and Cirencester. There was also some freight traffic offering in the shape of agricultural items and locally mined limestone from quarries which were rail connected. This view taken on 19th September 1959 reveals that a stone building which contained the usual amenities such as toilets and a waiting room, a signalbox and a lamp room occupied the up platform. What appears to be a coal scuttle stands outside the corrugated iron building on the left, no doubt used to top up fires in the signalbox and waiting room. With just one scheduled passenger train in each direction daily to deal with after the 1958 timetable changes the signalman's life here was uneventful to put it mildly. *LRF4386*

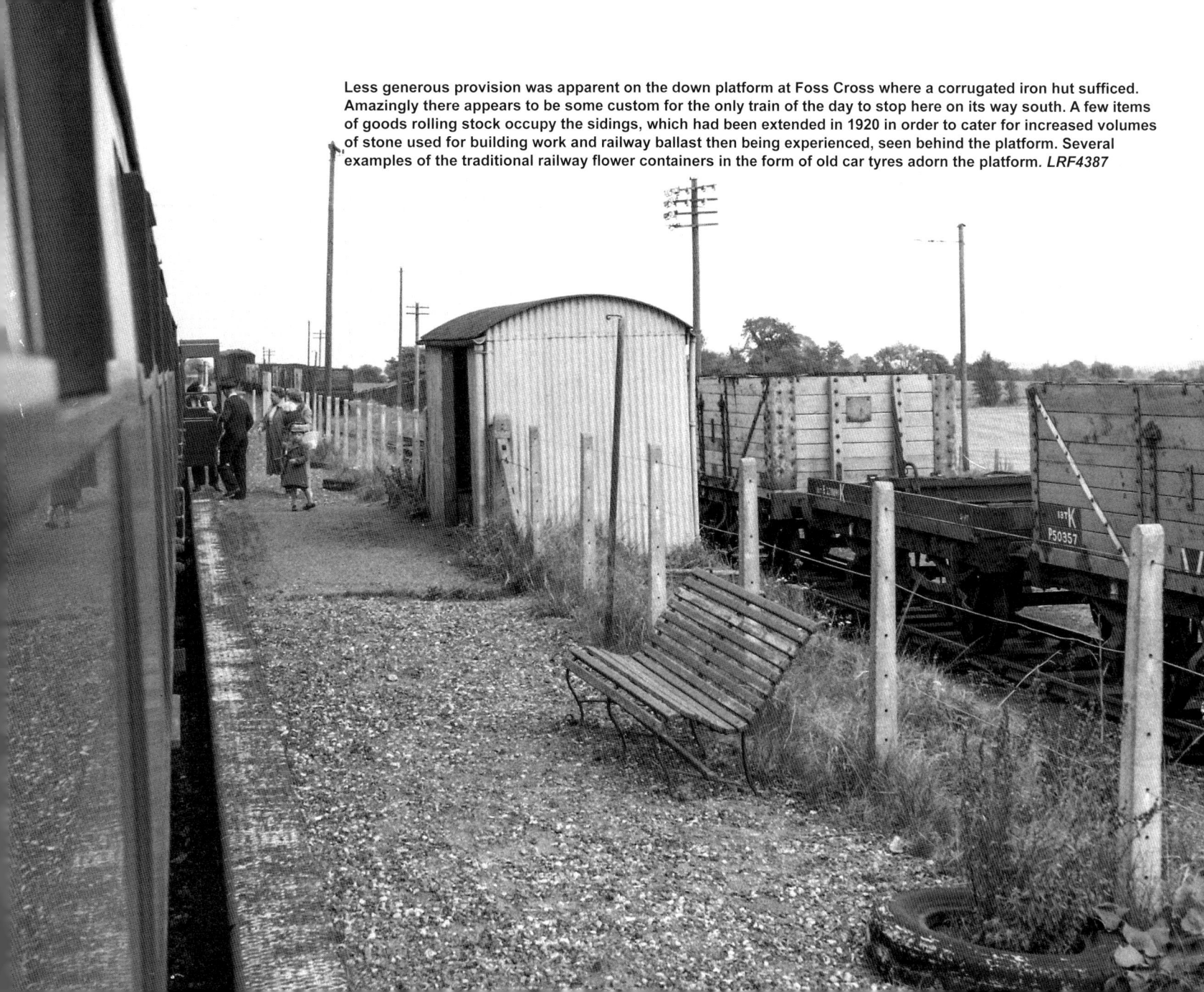

Less generous provision was apparent on the down platform at Foss Cross where a corrugated iron hut sufficed. Amazingly there appears to be some custom for the only train of the day to stop here on its way south. A few items of goods rolling stock occupy the sidings, which had been extended in 1920 in order to cater for increased volumes of stone used for building work and railway ballast then being experienced, seen behind the platform. Several examples of the traditional railway flower containers in the form of old car tyres adorn the platform. *LRF4387*

The delightful station at Chedworth, which became an unstaffed halt in February 1954, situated in the equally delightful village of Cotswold stone houses is seen from the Cheltenham – Southampton through service on 15th August 1959. A row of three fire buckets and an ornate lamp hanging under the canopy add charm to this view as does the old School road sign seen on the right. Today property in Chedworth is very sought after and no doubt a railway offering a service into Cheltenham would have boosted prices even more. *H960*

Looking north as the guard gives the driver the "right away" a southbound service departs Withington on 19th May 1959. Following the singling of this section of the route in 1928 a loop was retained here until 1957 but by the time of this view the former up platform was used by services in both directions. Although not appearing on the running in board the suffix "Glos" was added to the timetable description of Withington in 1924 to distinguish it from a similarly named intermediate station on the Ledbury to Hereford line which coincidentally also closed in 1961. *LRF4384*

The disused down platform at Withington is already starting to return to nature in this May 1959 view. The signalbox which contained 14 levers was in use until May 1956 at which time staffing at the station ceased and goods facilities were withdrawn. In November 1957 the box was converted to a ground frame. Track recovery took place here in 1963 with steam returning in the form of a 2800 class 2-8-0 hauling the demolition train. *LRF4383*

Andoversford & Dowdeswell station, the platform of which is all that remains in this view, dated from 1891. Double track extended northwards from the opening date but southwards remained single until doubling took place in 1900 only to be followed in 1928 by reversion to single track. Passenger trains ceased to stop after April 1927 but the goods yard here, which was accessed via the spur seen on the left which diverged from the former up line which was operated as a long siding from Andoversford junction half a mile to the north, remained open as late as October 1962. After closure to passengers the station building served as a "Cafe – Open Day & Night" to travellers on the adjacent A40. *JJ1325*

Not for much longer would the Andoversford Junction signalman be handing over the token for trains on the M&SWJR as this view dates from 9th September 1961 the last Saturday of operation. The lines off to the far left led to Kingham the route from there to Cheltenham remaining open to passengers until October 1962 whilst the two grass grown sidings in the middle were exchange sidings for the two railways which, sometimes acrimoniously, met here. *H3192*

The WR station at Andoversford, which refused to handle **M&SWJR** services until 1904, is captured on 19th September 1959 looking west towards Cheltenham. The solitary **M&SWJR** departure in each direction here was supplemented by some half dozen WR services each way between Cheltenham and Kingham. *LRF4375*

The Southampton to Cheltenham Train

The most impressive working of the day over the M&SWJR was the through train from Southampton Terminus to Cheltenham Spa. The up working Is seen here getting away from the Ludgershall stop on the 9th July 1956 hauled by U Class No. 31629. SR motive power had been introduced to the line in 1953 and these Maunsell moguls, based at Eastleigh shed, achieved a monopoly on the through service during the late 1950s, the two or three coach loads proving no problem for them over this switchback route. At this time Lansdown station in Cheltenham was still the traditional terminus of this train but from November 1958 they were diverted to the WR terminus of St. James which rendered interchange with services to the Midlands and the North more difficult for through passengers. *R7511*

U Class No. 31613 departs Marlborough with the Southampton service on an unrecorded date at 3:48 pm, if on time, having just crossed with a northbound working from Andover to Swindon Town due away at 3:49 pm after a seven minute stop at this Wiltshire market town. Note the wagon on the right in the former horse dock siding, grain traffic continuing after closure to passengers when up to four goods trains per week plied the route northwards from Savernake until 1964. *H1812*

Class N variants of the Maunsell moguls also handled the Cheltenham – Southampton through train and an example is seen here at Marlborough on 26th July 1961 as No. 31818 waits at the platform. Somewhat optimistically two station staff with barrows were apparently required to deal with parcels traffic although, as usual, passengers were rather thin on the ground. It will have taken 110 minutes to run the 47 miles from Cheltenham to Marlborough with any travellers intrepid enough to make the whole trip having to endure a similar amount of time for the remaining 50 odd miles before journey's end at Southampton. *H2827*

A better patronised up Cheltenham service pauses at Swindon Town on 13th August 1955 with U Class No. 31629 at its head. The roofless state of the former M&SWJR headquarters seen on the right was remedied in later years through having a new roof fitted and the building becoming used as offices by a firm of accountants. The edge of the turntable, which somewhat unusually was situated rather remotely from the former locomotive shed, can be seen on the far left. *PY10030S*

Another view of the through train, this time Southampton bound, sees U Class No. 31793 in charge on 15th August 1959. The opportunity is being taken to replenish the 3,500 gallon tender from the platform mounted water crane. This example of Maunsell's moguls was rebuilt from one of his original River Class 2-6-4Ts named "River Ouse" whose working life as a tank locomotive was cut short by the Sevenoaks accident when a fellow class member "River Cray" derailed at speed necessitating a drastic rethink and ultimate conversion of the class to tender locomotives. *H965*

Earlier on the same day No. 31793 pauses at Andoversford with the Southampton service. Other examples of these useful 2-6-0s which made frequent appearances over the route were 31613, 31618, 31620, 31626, 31629, 31639, 31791, 31794, 31801, 31804, 31808, 31809, 31816 and 31818. A 71A Eastleigh shedplate is carried by 31793 which lasted in service until May 1964 some 36 years after its rebuild in 1928. *H959*

No. 31808 leaves the shed at Cheltenham Malvern Road to back down to St James station just ½ mile away on the final Friday, 8th September 1961, in order to haul the penultimate through service to Southampton. The depot here closed in March 1964 with Malvern Road station closing to passengers in January 1966. In the distance can be seen East signalbox a GWR Type D box dating from 1906. The "Honeybourne Line" cycle and footpath now occupies the trackbed here. *H3183*

Swindon HQ

By the date of this view, July 1960, the roof of "The Croft" the former M&SWJR headquarters has clearly been repaired and, as previously mentioned, was latterly occupied by accountancy firm Haines Watts. The firm is still in business today in Old Station House, Old Station Road, although now their offices look out onto an industrial estate. The building was occupied by the railway company from opening in 1881 until it ceased to be used as a headquarters after 1924. It was then taken over by the S&T department for a number of years. *NF046-30*

Prior to handing over the single line tablet for the onward journey to Swindon's main station the signalman at Swindon Town has a chat with the crew of 5700 Class pannier tank No. 9721 on 3rd September 1960. On Mondays - Fridays the connecting service between the two Swindon stations ran once daily, primarily for staff at Swindon Works, the departure from Town station being at 7:08 am with a return from the main station at 5:55 pm. On Saturdays there were two trains from the main station with two return workings from the Town. As the 3rd September 1960 was a Saturday the train seen here was almost certainly the 2:04 pm working, which had originated from Savernake Low Level, and which would arrive at the main Swindon station 10 minutes later. *H1797*

The exterior of Swindon Town station is seen on the final day, 10th September 1961, with a Ford Anglia 105E to the fore amongst the parked vehicles of the period. Like Marlborough, the refreshment room at Swindon Town remained open after closure to passengers not finally shutting its doors until 1st February 1965 when the well known BBC reporter Tom Salmon interviewed some of the regulars. The interior of the bar had remained unchanged for decades with Victorian carved beer handles, potted palms, gas lighting and a venerable till worthy of Arkwright's Stores, the whole enterprise having been run by Reg Townsend and his wife since 1933. Following his death the bar continued to be operated by his widow until closure. *LRF6585*

This shot gives us an opportunity to view the 55 foot turntable manufactured by Cowans Sheldon & Co. of Carlisle and installed here during 1904/5 when major improvements to facilities were undertaken at the Town station. The table was situated at the north western end of the station and in fact was the second to be provided here the original one having been positioned adjacent to the former two road locomotive shed which had been constructed at the opposite end of the site. Parked out of the way on the far side of the turntable is permanent way vehicle No. PWM 3078. *H1803*

Rounding the curve and passing under Devizes Road bridge comes an unidentified Maunsell mogul with the Cheltenham – Southampton service on 26th July 1961. Judging by the scaffolding visible under the bridge there would appear to have been a problem with the structure but it is still standing today as is the mock half timbered public house "The Plough" seen in the background. Originally built as a cottage in 1838 it had, by 1867, been taken over by Arkells brewery who were persuaded to sell part of the garden to accommodate the railway when it arrived in 1881.
H2824

A valedictory shot of two of the railway staff based at Swindon Town, Mr Webb the Station Master and booking clerk Mrs Williams, who pose in front of pannier tank No. 4697 on 8th September 1961 shortly before they would be made redundant. No doubt finding it difficult in the circumstances to raise a smile for the cameraman, this brings home the stark reality of railway closures which affected not just passengers but also the staff who had loyally served the line often for many years. *H3204*

Our final pair of images of Swindon Town harks back some 90 years to more prosperous times for the line. Both views date from February 1931 with 1600 Class pannier tank No. 1620 hauling a mixed rake and an unidentified 4500 Class with a lengthy goods service. *ICA GW331/ICA GW389*

Savernake Shuttle

Attracting some interest from a canine spectator, 5700 Class pannier tank No. 9672 has charge of the shuttle service to Savernake Low Level at Marlborough on 15th March 1958. Constructed in 1948 this Collett designed locomotive was completed at Swindon Works under BR auspices and had a service life of 17½ years before withdrawal at the end of 1965. Of the five services between these points on Mondays – Fridays and six on Saturdays just two had their origin at Marlborough, the others starting from further afield. The slopes of Postern Hill, part of Savernake Forest, can be seen in the right background.
RCR11522

Replenishment time for 5700 Class pannier tank No. 3684 at Marlborough prior to taking out the 4:03 pm service to Savernake Low Level on 19th September 1959. This train provided a connection into a stopping service to London Paddington, arriving in the capital just before 7pm, and into a service for Bath and Bristol via Devizes. Three children on the far left obviously find the watering procedure a source of fascination and indeed at least one of them was invited onto the footplate in a subsequent image to be found in the photographer's collection. *H1816*

Arriving at Savernake with the aforementioned service No. 3684 will subsquently propel its stock out of the bay platform in order to run round and form the next departure, the 5:33 pm, for Marlborough. The carriages of the two coach "B" set dating from 1933 were recorded as W6359W/W6361W. *H1821*

The next shot in this sequence shows the train ready to depart from the bay and gives us a chance to study the layout at the western end of Savernake station. The signalbox, one of no less than eight that once served the immediate area, is Savernake West. The tall concrete posts on the platforms were used to hold the somewhat lethal paraffin vapour Tilley lamps which would be hoisted up the post after having been somewhat nervously lit by a member of platform staff. The old line from Savernake High Level and an overbridge can just be made out in the right distance behind the fence posts on top of the bay platform cutting side. *LRF4408*

The station forecourt at the Low Level station at Savernake plays host to a trio of typical vehicles of the period whilst one of the posters on the attractive stone station building extols the attractions of a trip to London. Closure of many of the intermediate stations between Westbury and Newbury took place in April 1966 and sadly Savernake was one of these casualties. The delights of "The Stores" at Burbage, the next village to the south, are advertised on the side of the van. *LRF4407*

Our final view of Savernake dates from 23rd October 1959 when, at just after 2pm according to the photographer's notes, 9F No. 92203 clatters through with an up goods service. At this time the 2-10-0, which was only some six months old, was allocated to St. Philips Marsh depot in Bristol and as is well known it went on to a life in preservation thanks to the artist David Shepherd who named it "Black Prince". Today it can be found on the North Norfolk Railway. *H1134*

Motive Power

Not often seen hauling services on the main line, Ivatt tanks were sometimes used on the Tidworth branch and such a service is seen here departing Ludgershall on 14th May 1955. No. 41305 is leaving from the main up platform rather than the bay and will shortly pass the impressive four arm signal gantry situated at the north end of the station. Services to Tidworth would come to an end in September of that year. Prior to the use of these LMS designed tank locomotives GWR prairie tanks of the 4500 Class were the usual motive power although the last train, on the 17th September, was entrusted to 4300 Class 2-6-0 No. 5396. *RCR6105*

One of the aforementioned 4300 Class No. 6320 departs Collingbourne with an Andover service on 8th July 1956. It was fitting that the lengthy association of this class with the route should be recognised on the final weekend of services when Nos. 6395, 6327 and 5306 of this class were all in action on freight, passenger and special workings. Note the immaculate state of the permanent way and the absence of lineside vegetation that blights the view from so many lines today. *R7493*

Another class synonymous with the line were the Manors and here, at Swindon Town on 13th August 1958, No. 7810 "Draycott Manor" restarts its train en route to Andover. Other members of the class known to have worked over the line include 7824 "Iford Manor" and 7808 "Cookham Manor", the latter working a special on the final weekend. In the bay platform lurks 4500 Class No. 4573. *PY10028S*

No. 5509 of the Churchward 4500 Class of prairie tanks was a regular performer on the line and is seen here running round at Marlborough on 12th August 1959. Its 8½ year residence on Swindon shed would come to an end in a couple of months' time when it was transferred to Bristol Bath Road at the beginning of October only to be withdrawn from service from Truro at the end of 1961. *PY10112*

As we have already seen, Maunsell moguls were the preferred motive power for the longer distance trains and this Andover service seen at Marlborough in July 1960 has U Class No. 31795 at its head. The signalbox seen in the left background dates from 1933 when the lines between Marlborough and Savernake were re-modelled. *NF044-38*

Pannier tanks such as this example of the 5700 Class No. 4697 were more than adequate for the light passenger loadings experienced on the line. With its smokebox optimistically chalked "Bristolian – Second Part" a two coach Andover service prepares to leave Swindon Town in July 1960. *NF044-16*

A fine study in light and shade at Marlborough concludes this short look at the motive power on the line as a Maunsell mogul heads north on 8th September 1961 passing 5700 Class pannier tank No. 8793. Other locomotive types to grace the route were 2200 Class Collett 0-6-0s and Standard 4MT 4-6-0s. Even more exotic types were to be seen on specials on the sections of the route that remained open after passenger closure such as a Britannia (70020), Merchant Navy (35023) and in the preservation era a Hall (4930) and even an S&D 7F (53809). *H3168*

Finale

The final weekend was one of celebration tinged with sadness and this view of pannier tank No. 8783 decorated for its last duty over the M&SWJR the following day was captured at Swindon Town on the 8th September 1961. Adorned with an extravagant smokebox floral display and carrying a headboard "Last Workman" it certainly looked the part, even if the locomotive is rather grubby, as it posed by the signal gantry situated to the south of the goods shed opposite the 27 lever "B" signalbox. *H3201*

In much better external condition and looking a credit to the cleaners no doubt owing to the railtour duty it would undertake two days hence, is No. 5306 seen at Savernake Low Level with a freight at approximately 5:30 pm on 8th September 1961. Close inspection of a label on the first box van, shown in greater detail in the second shot reveals it to contain "Lowes Quality Dog & Cat Food" a firm still going strong today. At the other platform is pannier tank No. 4697 which was working through from Swindon Town to Andover. *H3207/H3208*

Opposite: In addition to two special trains run on the final day, 10th September 1961, there were the two scheduled departures from Swindon to Andover and return. Here we see Nos. 6395 and 6327 passing an RCTS special, the coaches of which can be seen in the down platform and whose train occupants, almost all of whom are kitted out in jackets and ties, are busily engaged in photographing the up train, at Ludgershall. The brace of 4300 Class 2-6-0s with four coaches in tow was scheduled to depart Ludgershall with the 12:38 pm service from Andover at 12:53pm. The RCTS special was booked to wait at Ludgershall from 12:52 to 13:02 thus if all were running to schedule we are able to time this shot with some accuracy. The first coach of the service train is of interest being apparently a former Great Northern Railway brake vehicle. *AEB5835*

The aforementioned RCTS tour commenced from Swindon WR and No. 5306 hauling eight coaches is seen here being fitted with the headboard. The tour then proceeded to take the connecting spur at Rushey Platt en route to Andover. After just half an hour the tour returned north traversing the whole route and reaching St James station at Cheltenham at about 5pm. It then returned south to Swindon via Stroud and Kemble having covered close on 150 miles during the day. *LRF6557*

Having arrived at Andover Junction the fireman is engaged in uncoupling the stock in order that No. 5306 may be released to turn on the turntable located between the two locomotive sheds located here, one being used by the SR which was to close in 1962 and the other used by WR which had closed in 1958. A blind eye was clearly being turned by authority to trespassing on the track in those halcyon days. *AEB5837*

Tour participants as well as local schoolchildren and their parents turn out at Cirencester to witness the passage of No. 5306. I wonder how many of those children, now pensioners, remember this day more than 60 years ago. *LRF6592*

The other last day special, sponsored by the SLS as its headboard indicates, was hauled by Manor Class No. 7808 "Cookham Manor" seen here at Ludgershall about to depart for Andover as No. 5306 on the RCTS special, receiving acknowledgements from the SLS tour participants, runs through the station without stopping. *LRF6573*

Top: The SLS special with its seven coach consist of Mark 1 coaches storms away from its Marlborough stop on the southbound leg of this railtour which had started from Snow Hill station in Birmingham and run via Stratford-upon-Avon, Cheltenham Malvern Road and the M&SWJR to Andover. Billed as the final train to run over the northern half of the line it diverted to Swindon Works siding on its northbound return to collect those tour participants who had opted to visit the Works for an extended tour. *JJ1319*

Below: A fond farewell to the M&SWJR as a trio of villagers turn out to wave one of the last trains through Collingbourne Ducis station on 10th September 1961. VALETE M&SWJR! *LRF6560*

Want more...?

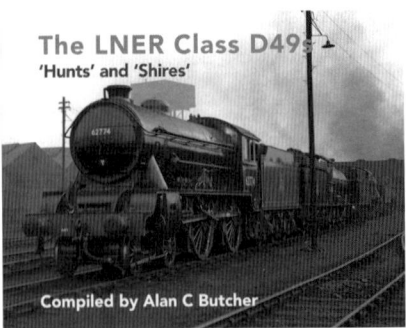

The LNER Class D49s by Alan C Butcher

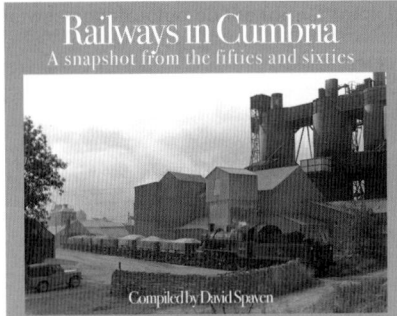

Railways in Cumbria - A Snapshot from the fifties and sixties by David Spaven

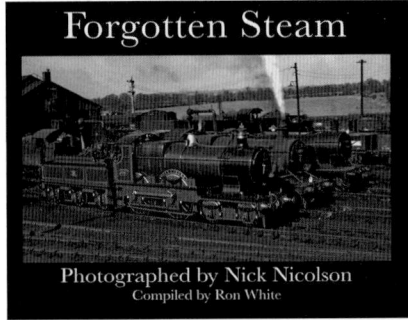

Forgotten Steam by Nick Nicolson

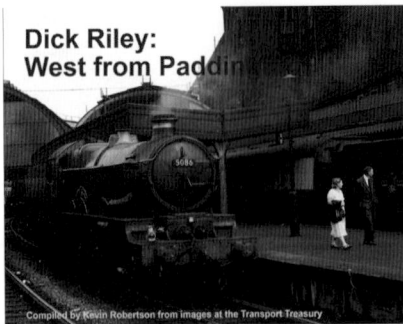

Dick Riley: West from Paddington by Kevin Robertson

Rails around Ireland - A Journey in Pictures by Charles P Friel and Michael McMahon

The LMS Jubilee Class based in Scotland 1935-1962 by Stuart Ashworth

Find us at ttpublishing.co.uk